Goodnight Gasoline: How to Win the Climate Change Challenge
by Twyla Dell Ph.D.

1st edition, 2021
ISBN 978-1-7335017-6-7
Edited and Adapted by Cory Sanchez
Book Design by Jessie Sanchez

Learn more at HowWeCanStopClimateChangeNow.com

20 lbs

That's the amount of CO_2 released into the air...

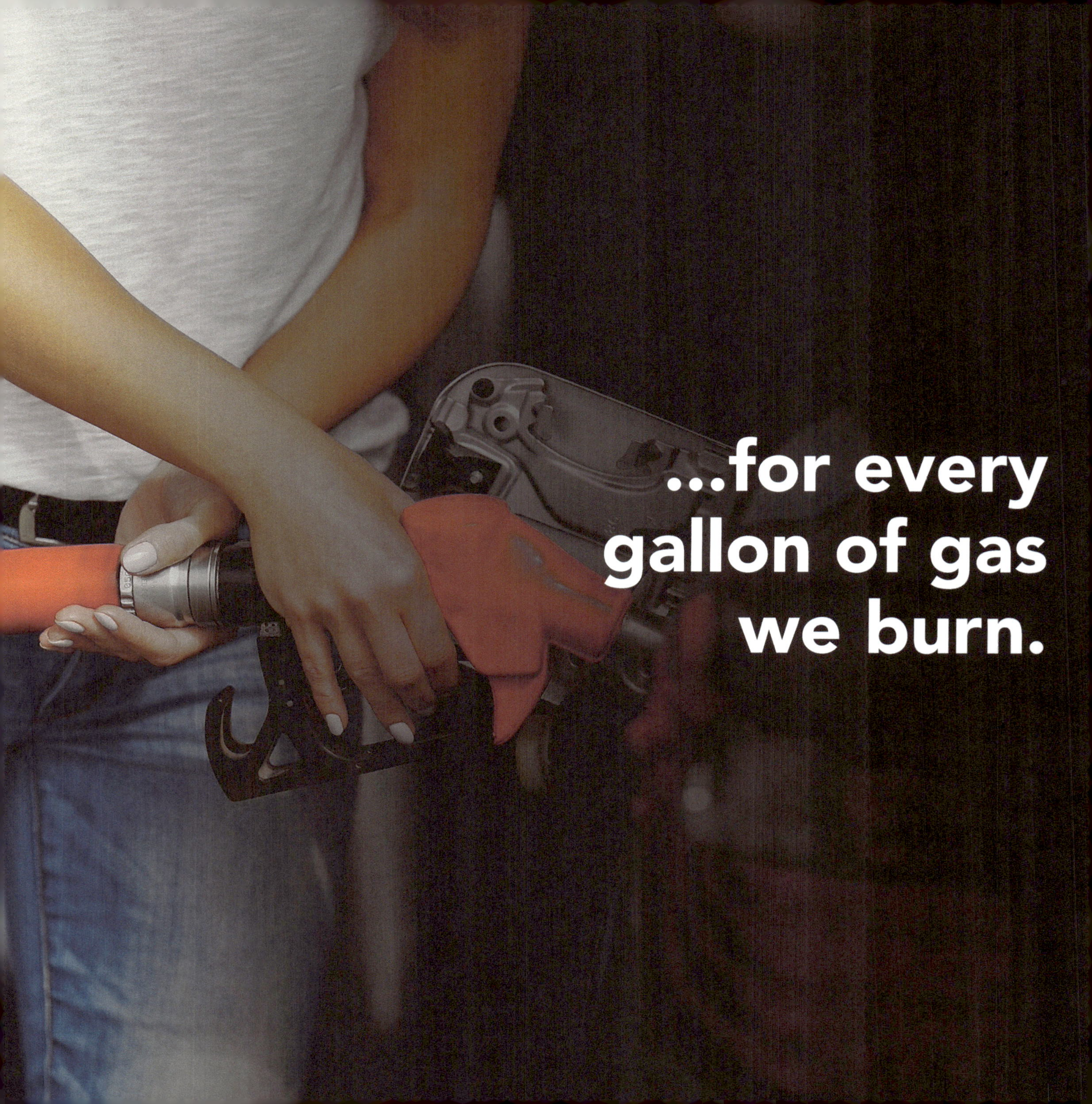
...for every
gallon of gas
we burn.

That's about
1 lb per mile.

How often do you drive or ride in a vehicle?

The average driver uses a full tank of gas per week.

A full tank of
gas averages

300 miles.

This adds up to 15,600 lbs of CO$_2$ per year per driver.

Every automobile contributes to this overwhelming number.

276 million vehicles fill their tanks and run on U.S. roads every day.

The world currently emits 51 billion tons of carbon dioxide a year.

This is enough to cause the sea level to rise...

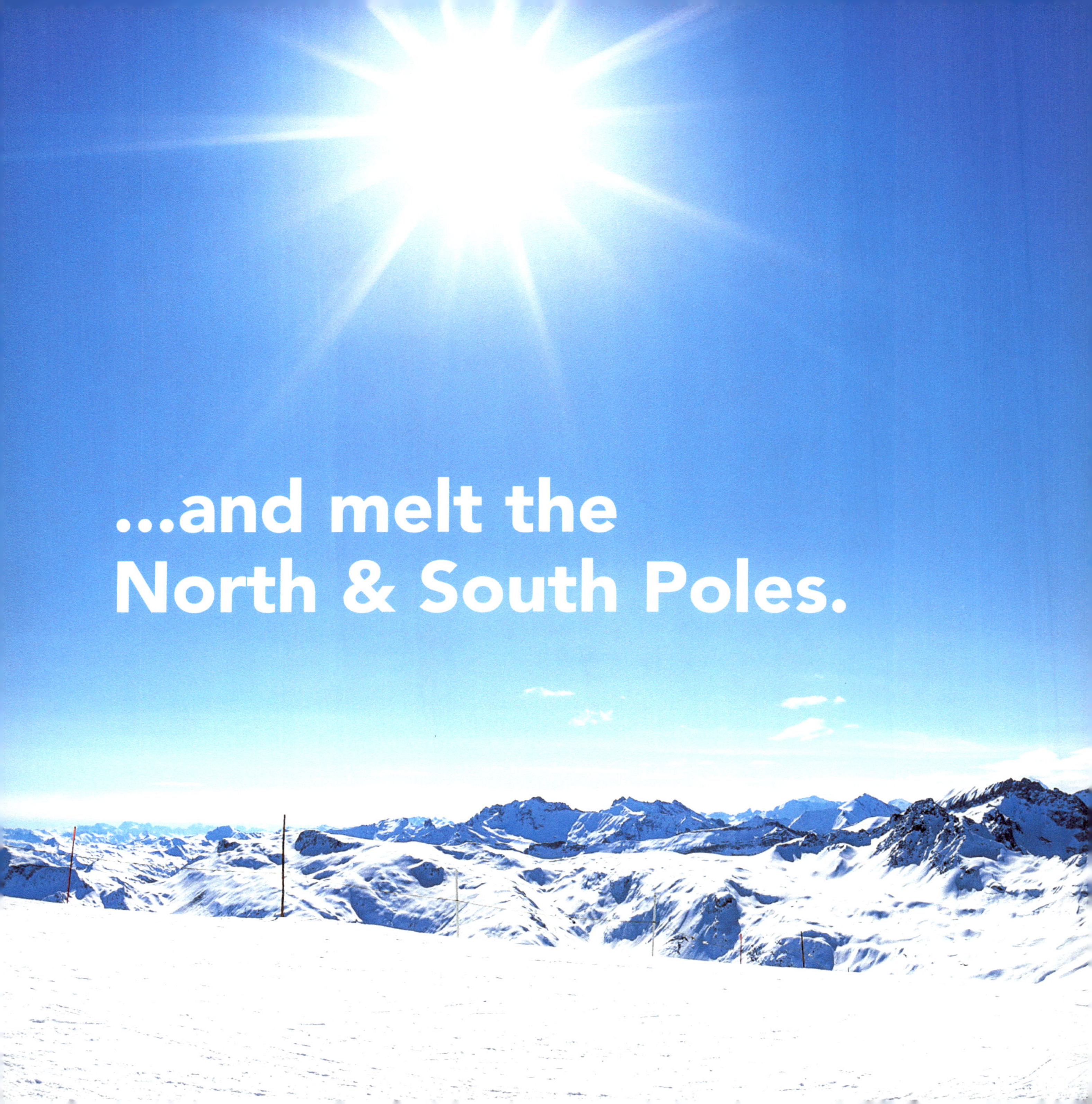

...and melt the
North & South Poles.

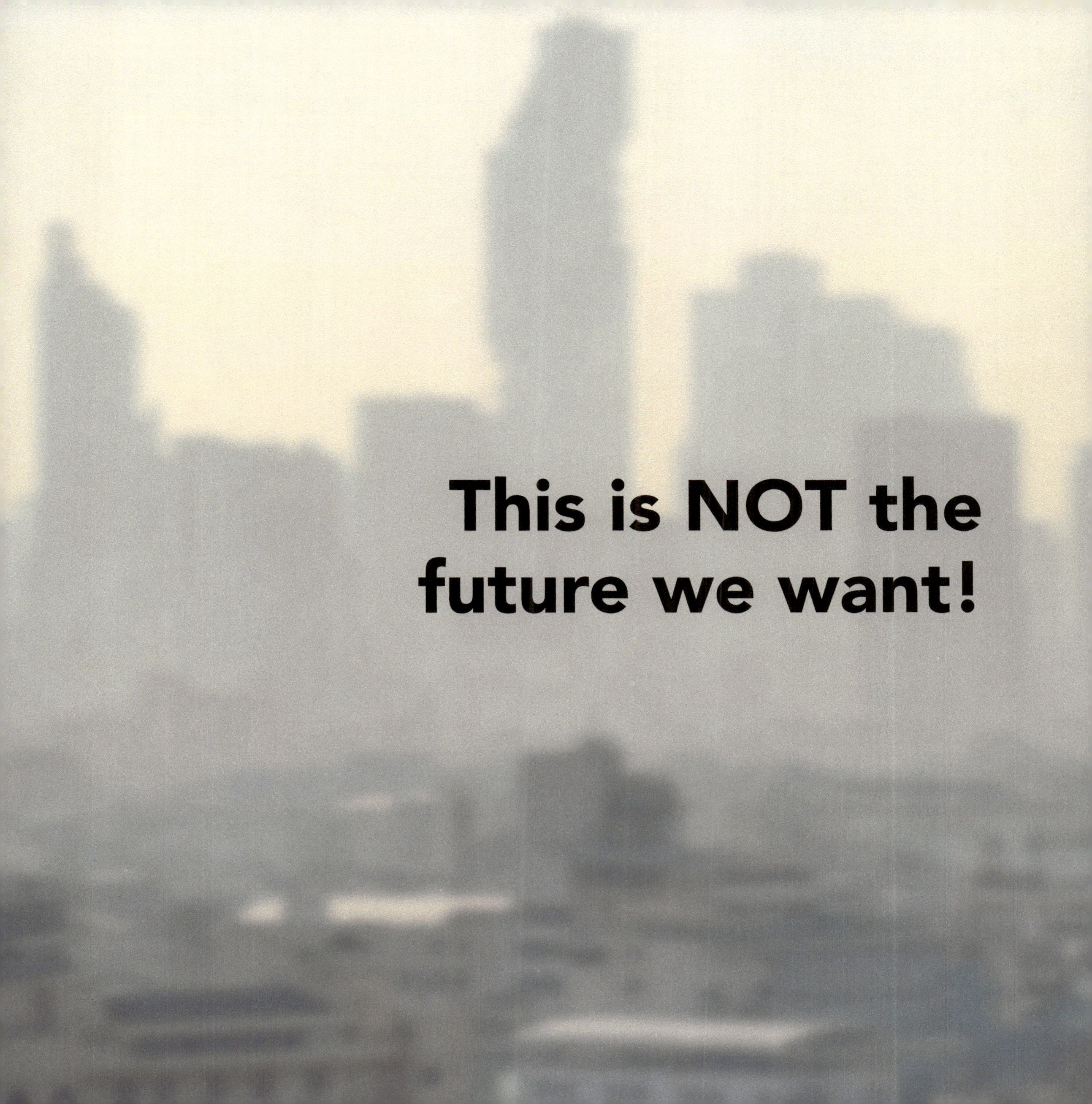

This is NOT the future we want!

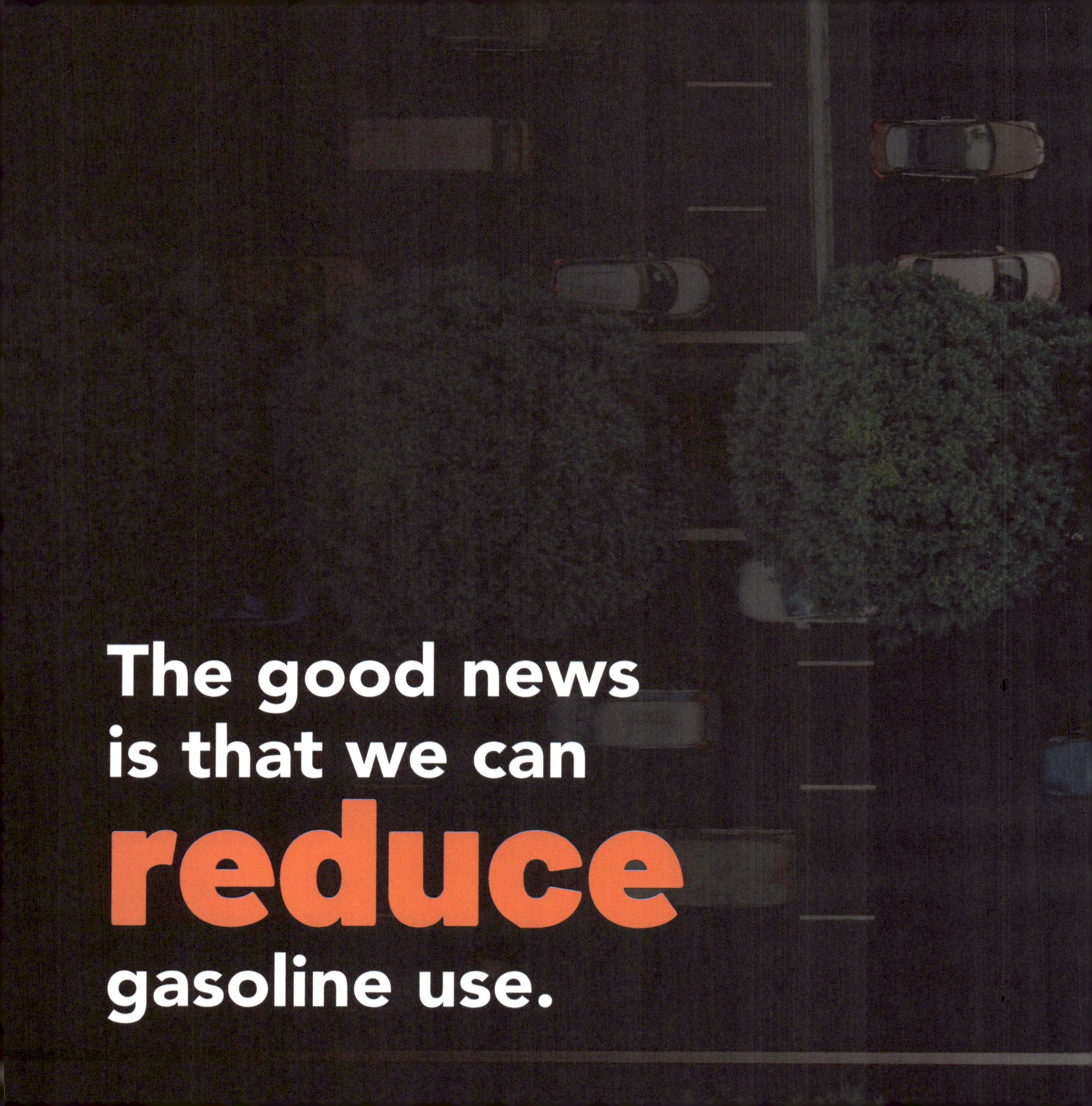

The good news is that we can **reduce** gasoline use.

There is no greater act of pollution for our generation than using gasoline.

There is no greater solution than to **stop burning gasoline.**

How can you help?

You can limit your use.

Keep track of your miles.

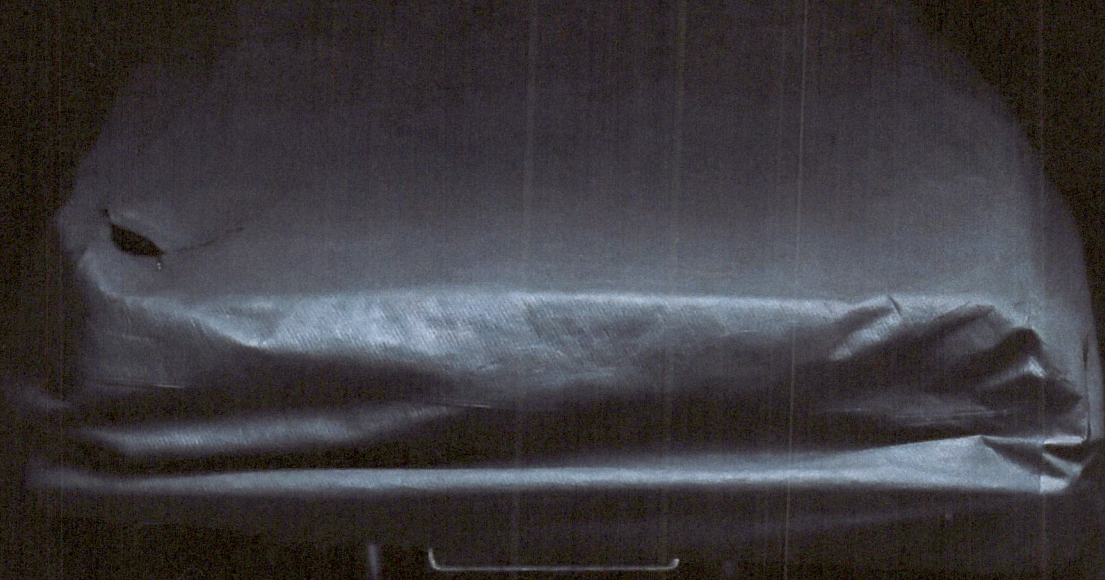

Drive less.

Walk more.

Ride a bicycle.

Use less

gasoline.

Switch from gas
to electric engines.

Spread the word.

Have discussions with your family and friends.

You can
make a difference.

Remember... There is no greater act of pollution for our generation than using gasoline.

There is
no greater
solution than
for people to...

...stop burning gasoline.

Please do your part.

You can make a difference.

I need your help.

Love,
Mother Earth

Twyla Dell is an author, environmental activist and fuel transition historian. Having written six environmental books, she is podcasting and blogging on reducing gasoline use. She worked for the EPA and received a Ph.D. in environmental studies from the Antioch University Graduate School.

Visit **HowWeCanStopClimateChangeNow.com** to see all of her blogs. Available for speaking engagements, webinars and is a seasoned podcast guest. Twyla may be reached at 913.209.0383.

Visit HowWeCanStopClimateChangeNow.com for more resources.

Sources:

Arenas, J. (2018). [Selective Focus Photography of Child's Hand]. Retrieved 2021, from https://www.pexels.com/photo/selective-focus-photography-of-child-s-hand-1250452/

CHUTTERSNAP. (2018). [Aerial shot of freeway traffic]. Retrieved 2021, from https://unsplash.com/photos/CATZGyxjzHk

Clifford, C. (2021, February 14). Bill Gates: These 5 concepts will help you understand the urgency of the climate crisis. Retrieved March 30, 2021, from https://www.cnbc.com/2021/02/14/bill-gates-concepts-to-understand-the-climate-crisis.html

Duncan, M. (2016). Highway 212, Lithonia, United States [Digital image]. Retrieved 2021, from https://unsplash.com/photos/IUY_3DvM__w

EnergySage, LLC. (2019, January 17). Pros and cons of electric cars. Retrieved March 30, 2021, from https://www.energysage.com/electric-vehicles/101/pros-and-cons-electric-cars/

Gates, Bill, HOW TO AVOID CLIMATE DISASTER: The Solutions We Have and the Breakthroughs We Need, Alfred A Knopf, New York, 2021, p. 3.

Hardy, M. (2018). Ocean Ripple [Digital image]. Retrieved 2021, from https://unsplash.com/photos/6ArTTluciuA

Haws, C. (2020). [Worn gas tank with disconnected hose]. Retrieved 2021, from https://unsplash.com/photos/2ATQm7PPhKQ

Hay, R. (2018). Oakland Bay Bridge, San Francisco, United States [Digital image]. Retrieved 2021, from https://unsplash.com/photos/PQzlMO1ifPA

Kemper, J. (2020). [Gas nozzle with cobwebs]. Retrieved 2021, from https://unsplash.com/photos/xwMQFszEMZc

Lindsey, R. (2020, August 14). Climate change: Atmospheric Carbon Dioxide: NOAA Climate.gov. Retrieved March 30, 2021, from https://www.climate.gov/news-features/understanding-climate/climate-change-atmospheric-carbon-dioxide

NASA. (n.d.). View of the Earth as seen by the Apollo 17 crew traveling toward the moon [Digital image]. Retrieved from https://unsplash.com/photos/vhSz50AaFAs

Nielsen, F. (2018). [Covered car]. Retrieved 2021, from https://unsplash.com/photos/eEpF3zclCfE

Preliminary data from *Petroleum Supply Monthly*. (2021, March 9). How much gasoline does the United States consume? Retrieved March 30, 2021, from https://www.eia.gov/tools/faqs/faq.php?id=23&t=10

Parjanen, V. (2020). [Grayscale Photo of a Gas Station]. Retrieved 2021, from https://www.pexels.com/photo/grayscale-photo-of-a-gas-station-3853870/

Perkins, P. (2020). [San Francisco 2020, after the labor day fires]. Retrieved 2021, from https://unsplash.com/photos/Z3_uSvERPfM

Ritchie, H., & Roser, M. (n.d.). Greenhouse gas emissions. Retrieved March 30, 2021, from https://ourworldindata.org/greenhouse-gas-emissions

September 2013 issue of National Geographic magazine. What the world would look like if all the ice melted. (2021, February 10). Retrieved March 30, 2021, from https://www.nationalgeographic.com/magazine/article/rising-seas-ice-melt-new-shoreline-maps#:~:text=If%20we%20keep%20burning%20fossil,sea%20level%20by%20216%20feet

Shvets, A. (2020). [Elderly woman with red hair in eyeglasses making video call]. Retrieved 2021, from https://www.pexels.com/photo/elderly-woman-with-red-hair-in-eyeglasses-making-video-call-5257218/

Statista Research Department. (2021, February 08). Number of cars in U.S. Retrieved March 30, 2021, from https://www.statista.com/statistics/183505/number-of-vehicles-in-the-united-states-since-1990/

How many registered motor vehicles are there in the U.S.? Some 276 million vehicles were registered here in 2019. The figures include passenger cars, motorcycles, trucks, buses, and other vehicles. The number of light trucks sold in the U.S. stood at 12 million units in 2019. Feb 8, 2021

Suter, B. (2019). Black Asphalt Road Near Mountains Under Cloudy Sky [Digital image]. Retrieved from https://www.pexels.com/photo/black-asphalt-road-near-mountains-under-cloudy-sky-3733269/

U.S. Energy Information Administration estimates. (2016, February 2). Carbon Dioxide Emissions Coefficients. Retrieved March 30, 2021, from https://www.eia.gov/environment/emissions/co2_vol_mass.php

Van Rooy, C. (2020). [Elephants in an African game reserve]. Retrieved 2021, from https://unsplash.com/photos/Y--zr3CPaPs

What the world would look like if all the ice melted. (2021, February 10). Retrieved March 30, 2021, from https://www.nationalgeographic.com/magazine/article/rising-seas-ice-melt-new-shoreline-maps#:~:text=If%20we%20keep%20burning%20fossil,sea%20level%20by%20216%20feet

Visit HowWeCanStopClimateChangeNow.com for more resources.

www.ingramcontent.com/pod-product-compliance
Lightning Source LLC
Chambersburg PA
CBHW060825270326
41931CB00002B/66

* 9 7 8 1 7 3 3 5 0 1 7 6 7 *